Diagonal Chromatic Number of a Maximal Planar Graph of

Diameter Three with Twelve Vertices

by

Stephanie Sierra

An Honors thesis submitted to the

Department of Mathematics in partial fulfillment of the

requirements for the Honors Program and the degree of Bachelor of Arts

Meredith College

Raleigh, NC

May 5, 2020

Student Name _Stephanie Sierra_ Date _5/5/2020_

Thesis Advisor _Jennifer Hontz_ Date _5/5/2020_

Honor's Director _Jennifer D. McMillen_ Date _5/6/2020_

Publication Agreement

I hereby grant to Meredith College the non-exclusive right to reproduce, and/or distribute this work in whole or in part worldwide, in any format or medium for non-commercial, academic purposes only.

Readers of this work have the right to use it for non-commercial, academic purposes as defined by the "fair use" doctrine of U.S. copyright law, so long as all attributions and copyright statements are retained.

Meredith College may keep more than one copy of this submission for purposes of security, backup and preservation.

Stephanie Sierra

May 5, 2020

Abstract

The diagonal chromatic number is the smallest number of colors needed to color the vertices of a graph such that any two adjacent or diagonal adjacent vertices are of different colors. This concept was first introduced by Bouchet in 1987. He showed that any plane graph, where the inner faces are triangles, is diagonally 12-colorable. However, in 1990, Borodin was able to show and reduce this bound to 11. In 1995, Sanders and Zhao furthermore reduced the bound to 10-colorable. A similar idea one can consider is how to construct a planar graph such that the diagonal chromatic number is less than 10 colors. In 2019, Danjun Huang, Yiquao Wang, Jing Lv, Yanping Yang, and Wefan Wang from Zhejiang Normal University in China focused on the structures of maximal planar graphs of diameter two and their diagonal coloring. They showed that if a graph is maximal planar with diameter two, then the diagonal chromatic number of the graph is nine if and only if the graph is a particular graph of nine vertices. This thesis is an exploration of the diagonal chromatic number of maximal planar graphs of diameter less than four. A maximal planar graph is a graph that consists of vertices and edges such that no new edge can be added without intersecting another edge. This thesis will aim to show that if a graph is maximal planar of diameter less than four with twelve vertices, then the diagonal chromatic number of the graph is six.

Table of Contents

Background 1

    Distance and Diameter 1

    Planarity 2

    Maximal Planar Graphs 3

    Four-Color Theorem 4

Structures 9

Diagonal Chromatic Number 22

Conclusion 42

Bibliography 44

Background

Graphs are utilized in various areas beyond the academia realm. Graphs are the basic abstraction for many real-world problems in today's society. They are used in day to day life such as in sales and marketing to find correlation between variables or using a Global Positioning System, GPS, to navigate from a location to a destination. Graphs are all around and their possibilities are endless.

In mathematics, graph theory is the study of graphs, which are mathematical structures that consists of both vertices and edges. Given a graph G, we usually use V(G), and E(G), to denote its vertex set, and edge set, respectively. Below is an illustration of a simple graph that consist of two vertices and one edge.

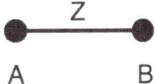

Figure 1:V(G) = {A, B}, E(G) = {Z}

Distance and Diameter

There are infinite ways to construct a graph; however, all graphs considered in this paper are constructed based on the distance and diameter between vertices. The distance between two vertices is the length of a shortest path connecting them in G. If there are multiple paths that connect the two vertices, then the shortest path is considered the distance. For example, in figure 2, there are three paths from vertex A to vertex B:

1. AB, which has a length of 1, since there is one edge between them.

2. AD, DB, which has a length of 2, since there are two edges between them.

3. AC, CD, DB, which has a length of 3, since there are three edges between them.

The shortest length from vertex A to vertex B is the direct path AB. The distance between vertex A and vertex B is one.

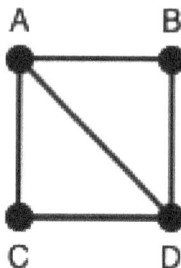

Figure 2: A four vertices graph with five edges.

The diameter of a graph G is the greatest distance between any two vertices. In figure 2, the distance between vertex B and vertex C is two, therefore the diameter of the graph is two.

Planarity

When constructing graphs, it's important to be able to draw graphs without crossing edges. A graph that can be drawn in the plane so that no two of its edges cross each other is called planar. Figure 3 is an example of isomorphic graphs; however, figure 3a is nonplanar because the edges intersect with other edges. Figure 3b is planar, because there aren't any edges that cross each other.

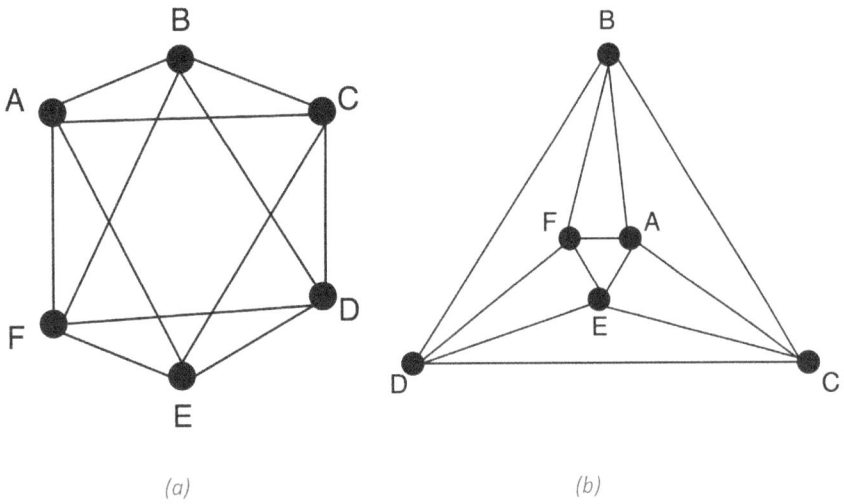

Figure 3: Nonplanar vs Planar

Maximal Planar Graphs

This paper will focus on the structure of maximal planar graphs. A maximal planar graph also known as plane triangulation is a graph in which every face is a triangle or bounded by three edges (Huang, Wang and Lv). A maximal planar graph must also follow planarity to which no new edges can be added without crossing another. Moreover, the following equation is helpful in testing maximal planarity of a graph (Huang, Wang and Lv).

$$|E(G)| = 3|V(G)| - 6$$

Examining Figure 2, it is noticeable that the graph is not maximal planar. The graph has five edges; however, by utilizing the equation above, the graph should have six edges in order to be considered as maximal planar. Thus, the addition of an edge from vertex B to vertex C is necessary to make Figure 2 a maximal planar graph. Figure 4 demonstrates the newly constructed graph.

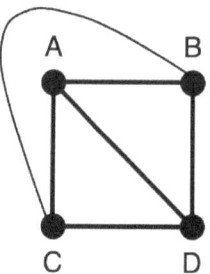

Figure 4: Maximal planar graph

Four-Color Theorem

In the domain of graph theory, an ongoing subtopic that is still very active in the field of research is graph coloring. Graph coloring is the assignment of colors to certain objects in a graph. Such objects can be vertices, or edges (Chartrand and Zhang). This topic was first introduced by Augustus De Morgan, a mathematics professor at the University College in London in 1852 (Chartrand and Zhang 261). He initiated the question that formed the Four-Color Theorem, which states that every planar map can be colored with at most four colors in such a way that neighboring countries/regions are colored differently (Chartrand and Zhang). This theorem is important because it led to many other mathematical revelations in vertex coloring that include the chromatic number of a graph. The chromatic number of a graph G, denoted by $\chi(G)$, is the smallest number of colors needed to color the vertices of a graph such that no two adjacent vertices have the same color (Chartrand and Zhang). The structure drawn in Figure 2 can be colored with three colors such that each adjacent vertex in the graph is assigned a different color.

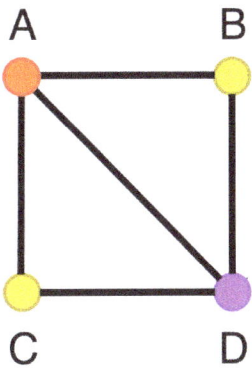

Figure 5: χ(G) = 3

There are many subtopics related to graph coloring and chromatic number of a graph. Bouchet et al. in 1987 introduced a similar concept to chromatic number known as the diagonal coloring of plane triangulations. The diagonal chromatic number is the smallest number of colors needed to color the vertices of a graph such that any two adjacent or diagonal adjacent vertices are of different colors (Huang, Wang and Lv). Coloring diagonal adjacent vertices are explained below.

How to find the Diagonal Chromatic Number of a graph

The diagonal chromatic number of a graph is found by coloring the vertices of a graph such that not only the adjacent vertices have different colors, but also the diagonal vertices must also have different colors. This will be shown by example with Figure 3b.

It is important to note that the beginning vertex can be selected at random. In this example, we will begin with coloring vertex A with the color green. This means that vertices B, C, E, and F cannot be green because those vertices are directly adjacent to A. In addition to those vertices, vertex D cannot be green because vertices A and D are both adjacent to vertices F and E, which of course share an edge. Next, we will color vertex F pink and vertex E purple. Figure 6 is an example of the vertices A, E and F with their assigned color.

Since the main objective is to find the smallest number of colors needed to color the graph, it is important to rule out the possibility of the remaining vertices being pink, green or purple before adding a new color to the graph.

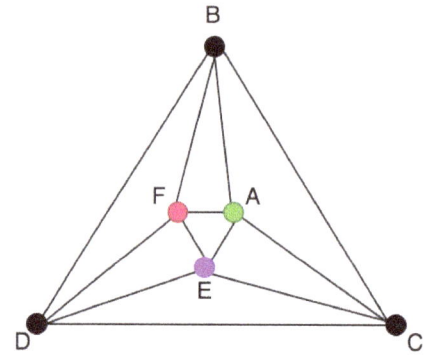

Figure 6: Diagonal coloring of vertices A, E and F

Vertex B cannot be colored pink or green because vertices F and A are adjacent to B. It also cannot be colored purple because vertex E, which is diagonally adjacent to B, is purple. This means that a new color needs to be added to the list of colors. Vertex B will be colored blue.

Next, we will look at vertex D. Vertex D cannot be colored blue, pink or purple because the vertices adjacent to D have been assigned with those colors. Vertex D also cannot be colored green because A is diagonally adjacent to D. This means that a new color must be added to the list of colors in order to color the vertex. In this example, vertex D will be orange. Pictured below are the five colors needed to color five out of the six vertices in the graph.

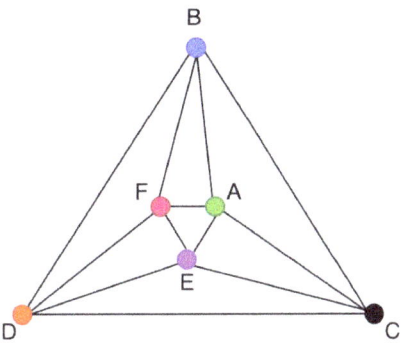

Figure 7: Diagonal coloring of vertices B and D

Lastly, vertex C cannot be assigned with the colors blue, green, purple or orange due to the fact that the vertices adjacent to C have been assigned with those colors. In addition to those colors, vertex C cannot be colored pink because it is diagonally adjacent to F. This means that vertex C must be given a color that has not been assigned to a vertex yet. Vertex C will be yellow.

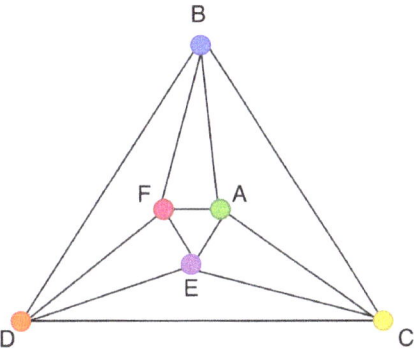

Figure 8:Diagonal coloring of vertex C

The diagonal chromatic number of the graph is six. How is the diagonal chromatic number different from the chromatic number of a graph? Let's take a look. Using the same graph, we will determine the chromatic number. First, we will assign vertex A with the color green, F with pink and E with purple exactly like figure 6. Then we will look at the remaining vertices.

Vertex B cannot be pink or green; however, it can be purple because vertex E is not adjacent to B. So, vertex B will be assigned with the color purple. Next we'll look at vertex D. Vertex D cannot be pink, or purple, but it can be green because vertex A is not adjacent to D.

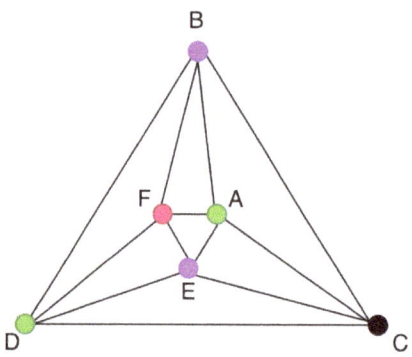

Figure 9: Chromatic coloring of vertices B and D

Lastly, vertex C cannot be green or purple, but it can be assigned with the color pink.

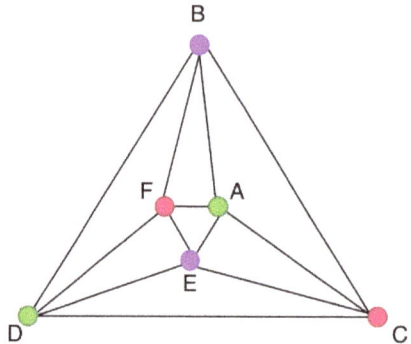

Figure 10: Chromatic coloring for vertices B, C and D

Since it only took three colors to color the graph such that no adjacent vertices were assigned the same color then this indicates that the chromatic number of the graph is three, $\chi(G) = 3$.

Bouchet et al. showed that any planar graph, where the inner faces are triangles, is diagonally 12-colorable. In 1990, Borodin was able to show and reduce this bound to 11. In 1995, Sanders and Zhao furthermore reduced the bound to 10-colorable (Huang, Wang and Lv).

A similar idea one can consider is how to construct a planar graph such that the diagonal chromatic number is less than 10 colors. In 2019, Danjun Huang, Yiquao Wang, Jing Lv, Yanping Yang, and Wefan Wang from Zhejiang Normal University in China focused on the structures of maximal planar graphs of diameter two and their diagonal coloring. A graph that is of diameter two has at least five vertices and the maximum distance between any two vertices is two. These researchers showed that if a graph is maximal planar with diameter two, then the diagonal chromatic number of the graph is nine if and only if the graph is a particular graph of nine vertices.

This thesis is an exploration of the diagonal chromatic number of maximal planar graphs of diameter less than four. This thesis will aim to show that if a graph is maximal planar of diameter less than four with twelve vertices, then the diagonal chromatic number of the graph is six.

Structures

This section is devoted to investigating the structures of maximal planar graphs of diameter three, which will be applied to the study of the diagonal chromatic number of graphs. There are three main classification of graphs that we will evaluate. Each of these graphs will have a diameter of at most three and a maximum of twelve vertices. For conciseness, a maximal planar graph of diameter three is called a MP3-graph. This discussion is based partially on the result of Huang, Wang and Lv.

The three main structures that we will evaluate are constructed around basic shapes. These figures are triangle, square, and pentagon.

Structure 1: The Triangle

An MP3-graph that begins with a triangular shape has many unique characteristics. In order to construct an MP3-graph with 12 vertices we'll need to begin with the basic figure of a triangle.

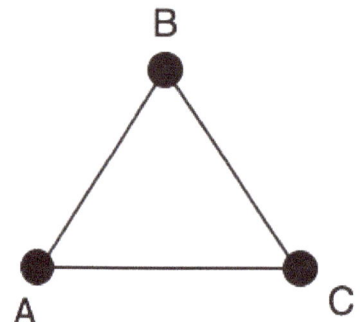

Figure 11: Base shape of the triangular figure

The next vertex drawn on the graph can be placed adjacent to any vertex on the triangle. In this example, we will place the fourth vertex, D, above B. Vertex D must share an edge with all of the vertices of the original triangle.

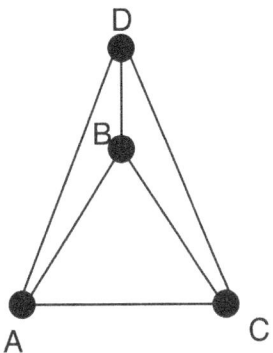

Figure 12: Expansion of base shape with an additional vertex D

The following vertex, E, can be placed adjacent to vertices A or C but not D. Vertex E must

share an edge with vertices A, C, and D. Vertices B and E cannot share and edge because if there

were an edge, the graph would not be considered as planar. The overall goal is to expand the

graph by placing vertices in a clockwise or counterclockwise direction.

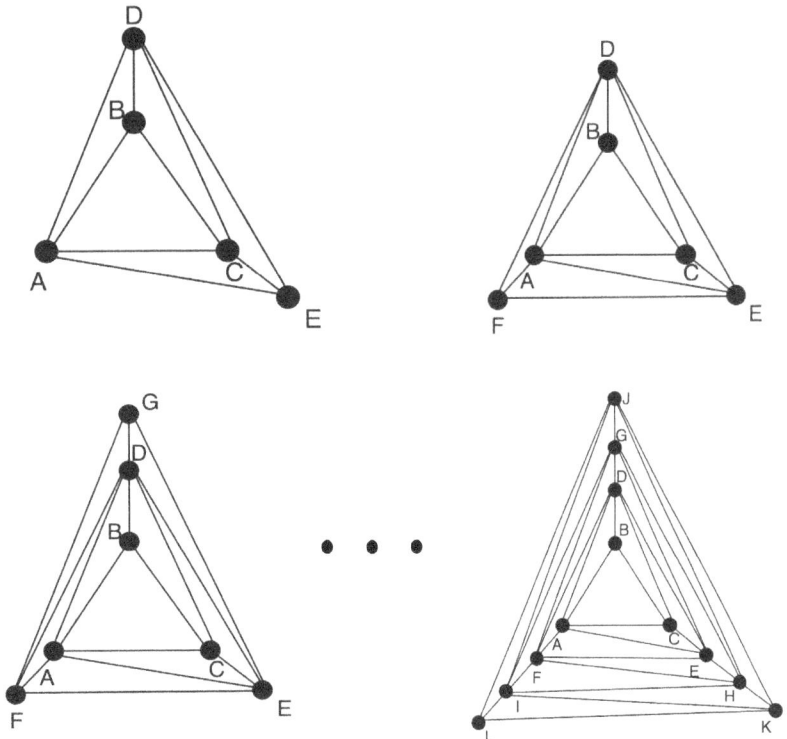

Figure 13: Expansion of the Triangle Structure

One of the many unique characteristics of this structure is that the initial shape, a triangle, is maximal planar, which means that no additional edges need to be added. This structure requires a maximum of three vertices to isolate the base structure from the rest of the graph. For example, removing vertices D, E, and F results in isolating triangle ΔABC from the rest of the graph.

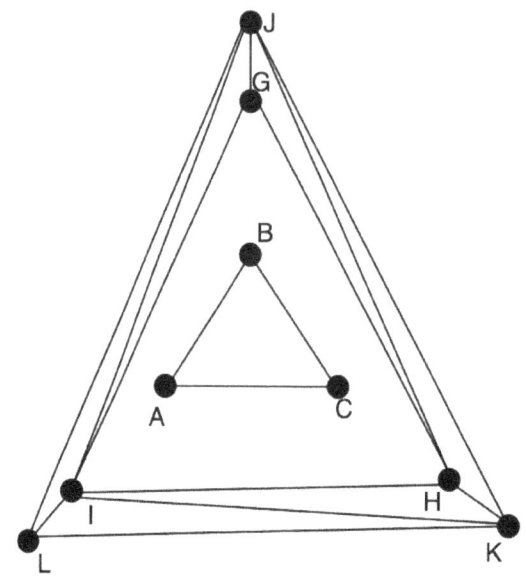

Figure 14: Isolating triangle ΔABC by removing vertices D, E and F

In addition to the foundational base of the structure being maximal planar, another unique characteristic of this structure is that it only requires eight vertices to classify as MP3-graph, which can be seen in the figure below.

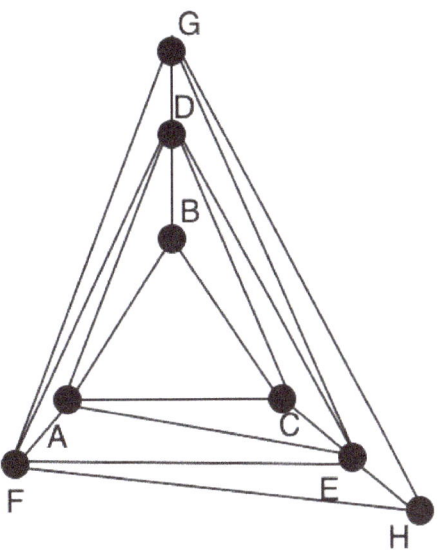

Figure 15: MP3-graph with only eight vertices

Structure 2: The Square

A square is a four-sided figure consisting of four vertices and four edges.

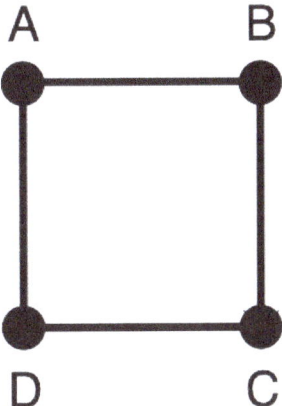

Figure 16: Base structure of the square structure

For our purposes, we will use a square as our base to develop an MP3-graph. We will first

convert the square into two triangles by drawing an edge that either connects vertices A and C or

vertices B and D. This figure still remains as a square with an additional edge across the center.

This edge is necessary because in order for this graph to be considered as maximal planar all inner faces of the graph must be bounded by three edges. One additional edge that connects

vertex B and D is needed to make this initial figure maximal planar.

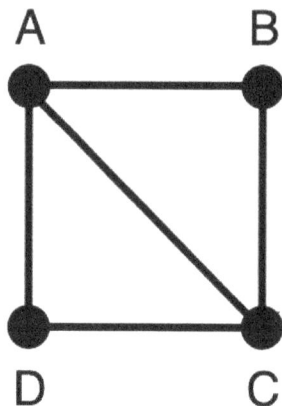

Figure 17: Converting the initial figure to be maximal planar

We will expand this figure in a similar way to the triangular structure. The next vertex E can be

adjacent to any of the vertices that are currently on the graph. In this example, we'll add vertex

E, next to vertex A. Vertex E will share an edge with A, B, and D. As you can see below, two

additional triangles have been added to the graph.

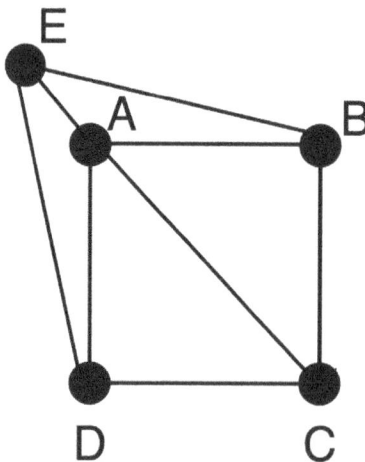

Figure 18: Expanding the initial base by adding vertex E

Vertex E does not share an edge with vertex C because without the edge that connects vertex E and B, then the inner faces of the graph would not all be triangles. Vertices A, B, C, and E make a square.

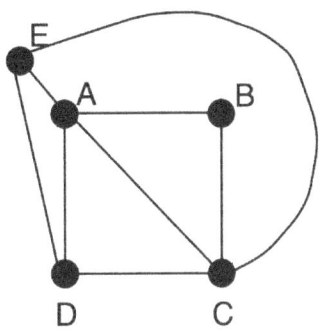

Figure 19: Incorrect way to expand graph

The next vertex F can be drawn adjacent to vertices B or D. In this example we will be expanding the graph counterclockwise, thus vertex F will be adjacent to D.

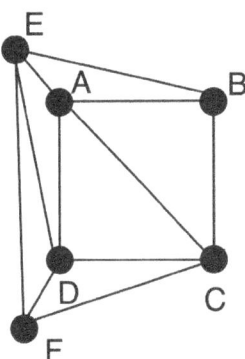

Figure 20: Square graph with six vertices

We will continue this pattern until the graph is of diameter at most three and has twelve vertices. This specific graph needs a minimum of twelve vertices to make a maximal planar graph of diameter three.

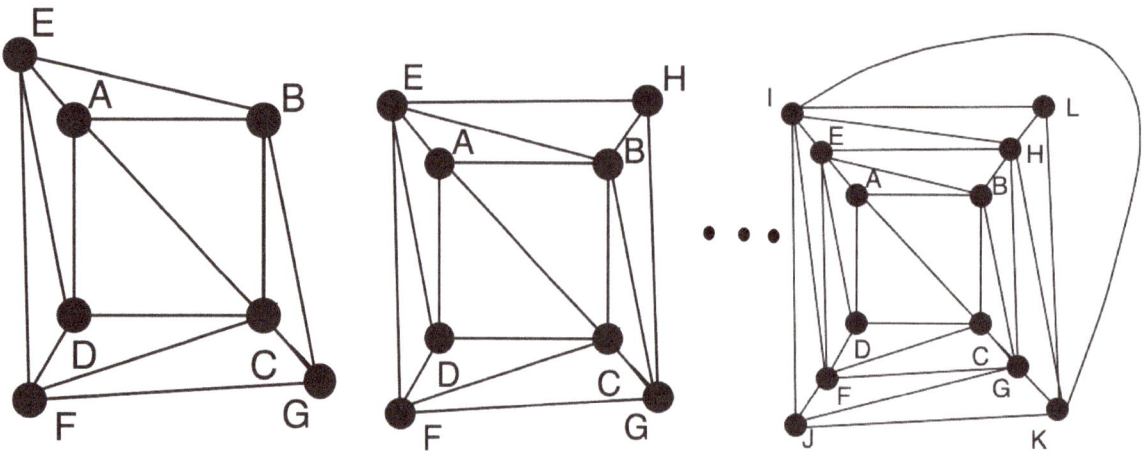

Figure 21: Expansion of graph until 12 vertices is reached

This structure may have similar expansion techniques as the triangular figure; however, the minimum number of vertices to isolate a triangle differs. The triangular figure requires a minimum of three vertices to be removed in order to isolate a triangle, while the square figure requires a minimum of four vertices to be removed in order to isolate a triangle. The graphs below are the resulting graphs after removing the necessary vertices from the square graph to isolate a triangle. A triangle cannot be isolated with the removal of exactly three vertices in the square graph. This implies that the triangle and square structures are not isomorphic.

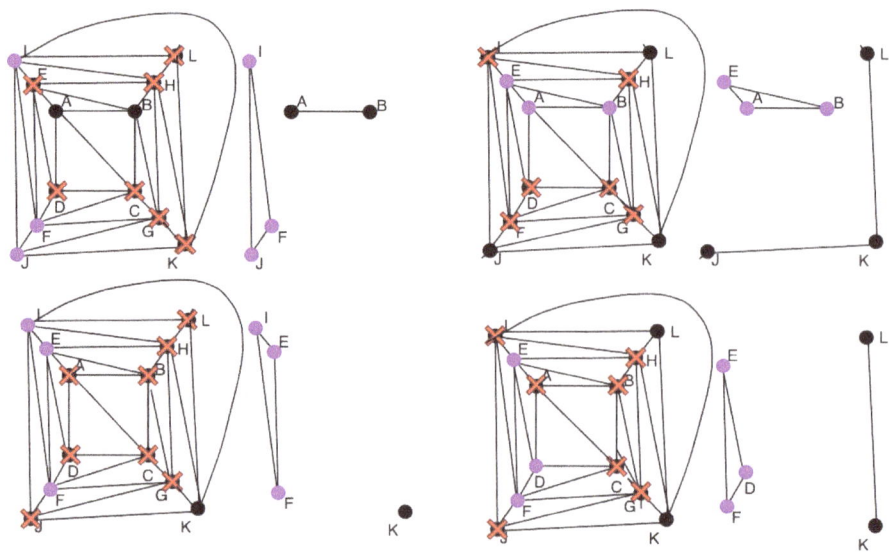

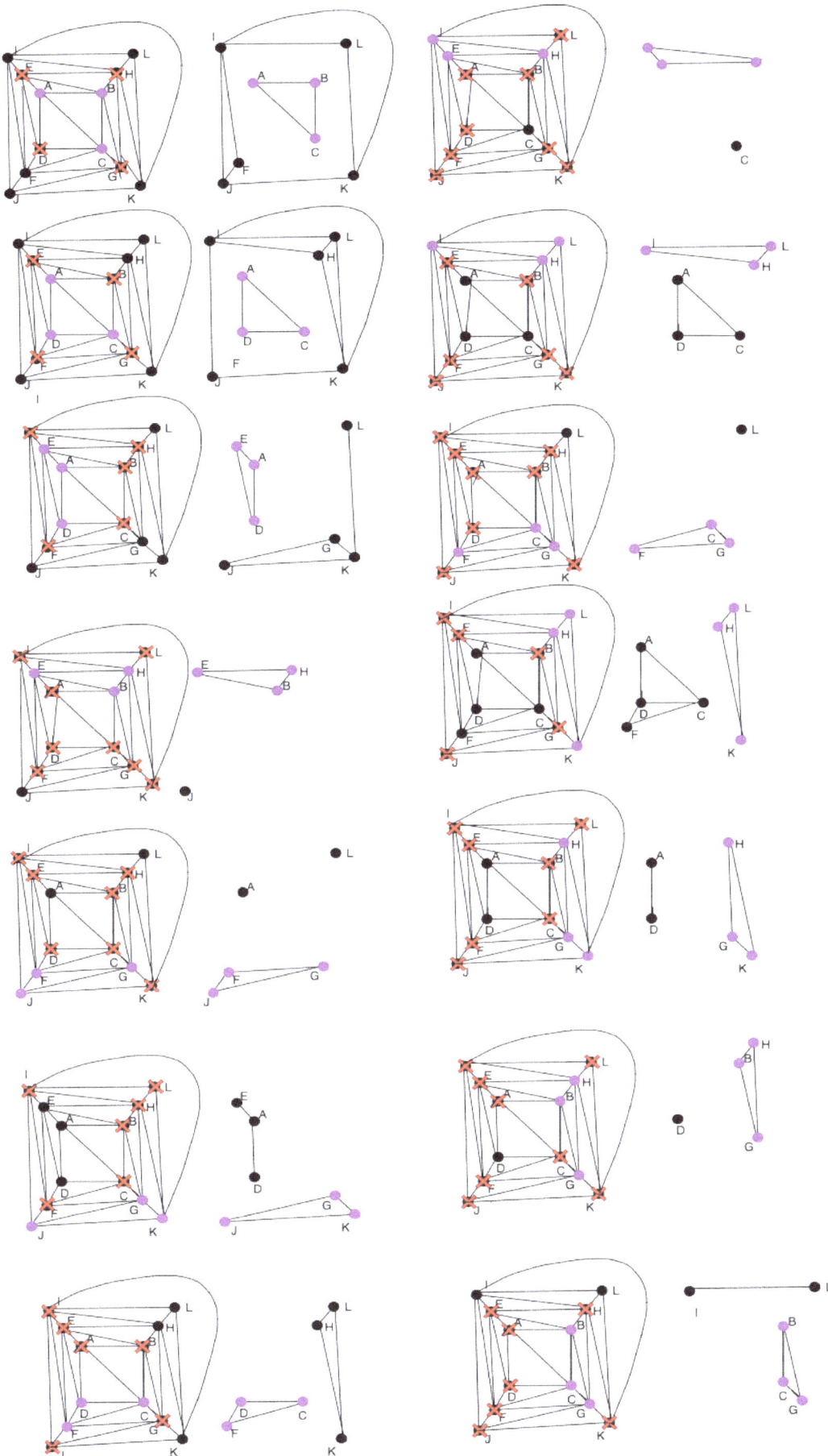

As shown above, a minimum of four vertices need to be removed in order to isolate a triangle in this figure.

Structure 3: The Pentagon

The final structure we will evaluate is a pentagonal MP3-graph. A pentagon is a polygon with five sides.

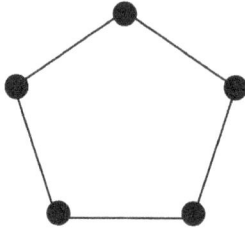

Figure 22: A pentagon

We will convert a pentagon into an MP3-graph by firstly transforming the polygon into three neighboring triangles by adding two edges. The edges must not intersect, which means that in development of this graph the two additional edges must share a vertex. In this case, vertex A will be the common vertex between the two edges.

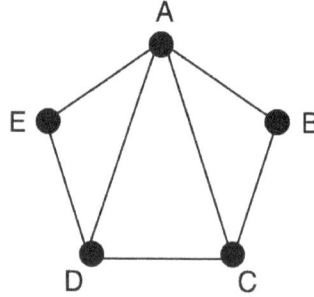

Figure 23: Converting the inital figure into a maximal planar graph

We will expand this figure in a similar way as the triangular and square structures explained above. The next vertex F can be adjacent to any of the vertices that are currently on the graph. In this example, we'll add vertex F adjacent to vertex A.

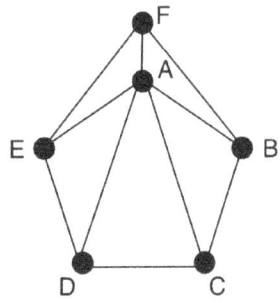

Figure 24: Expansion of base figure

Vertex F will share edges with the nearest vertices A, B, and E. As pictured above, two additional triangles have been added to the graph. It's important to note that vertex F cannot share an edge with vertex C or D because the inner faces of the graph would not all be triangles. If the inner faces of the graph are not all triangles, then the graph would violate the maximal planarity constraint.

The next vertex G can be drawn adjacent to vertices B or E. In this example we will be expanding the graph counterclockwise, thus vertex G will be adjacent to vertex E.

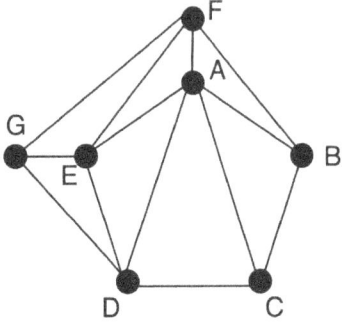

Figure 25: Initial figure with two additional vertices

We will continue this pattern until the graph is of diameter at most three and has twelve vertices. This specific graph is similar to the square graph because it also needs a minimum of twelve vertices to make a maximal planar graph of diameter three.

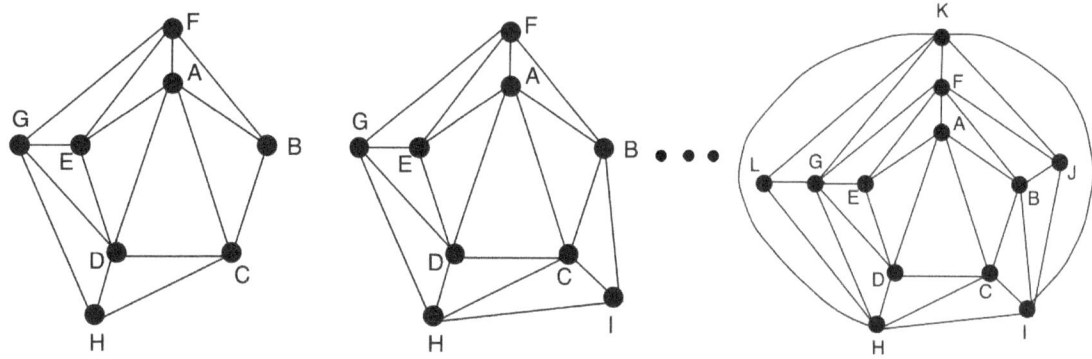

Figure 26: Expansion of the initial pentagon until 12 vertices is reached

The pentagonal structure requires a minimum of five vertices in order to isolate the base structure. To isolate the base structure, vertices F, G, H, I and J must be removed.

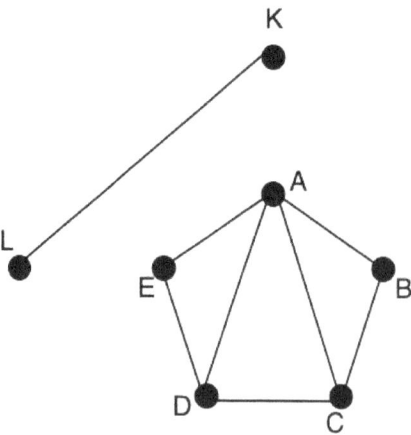

Figure 27: Isolating initial base shape required the removal of five vertices

This structure may have similar expansion techniques as the previous figures; however, the minimum number of vertices to isolate one triangle differs. A minimum of three vertices needs to be removed from the triangular structure in order to isolate a triangle; however, a minimum of four vertices need to be removed in order to isolate a triangle in the pentagonal structure. The following graphs will demonstrate that the triangle and pentagon structures are not isomorphic. Just as before, four vertices are needed to isolate a triangle.

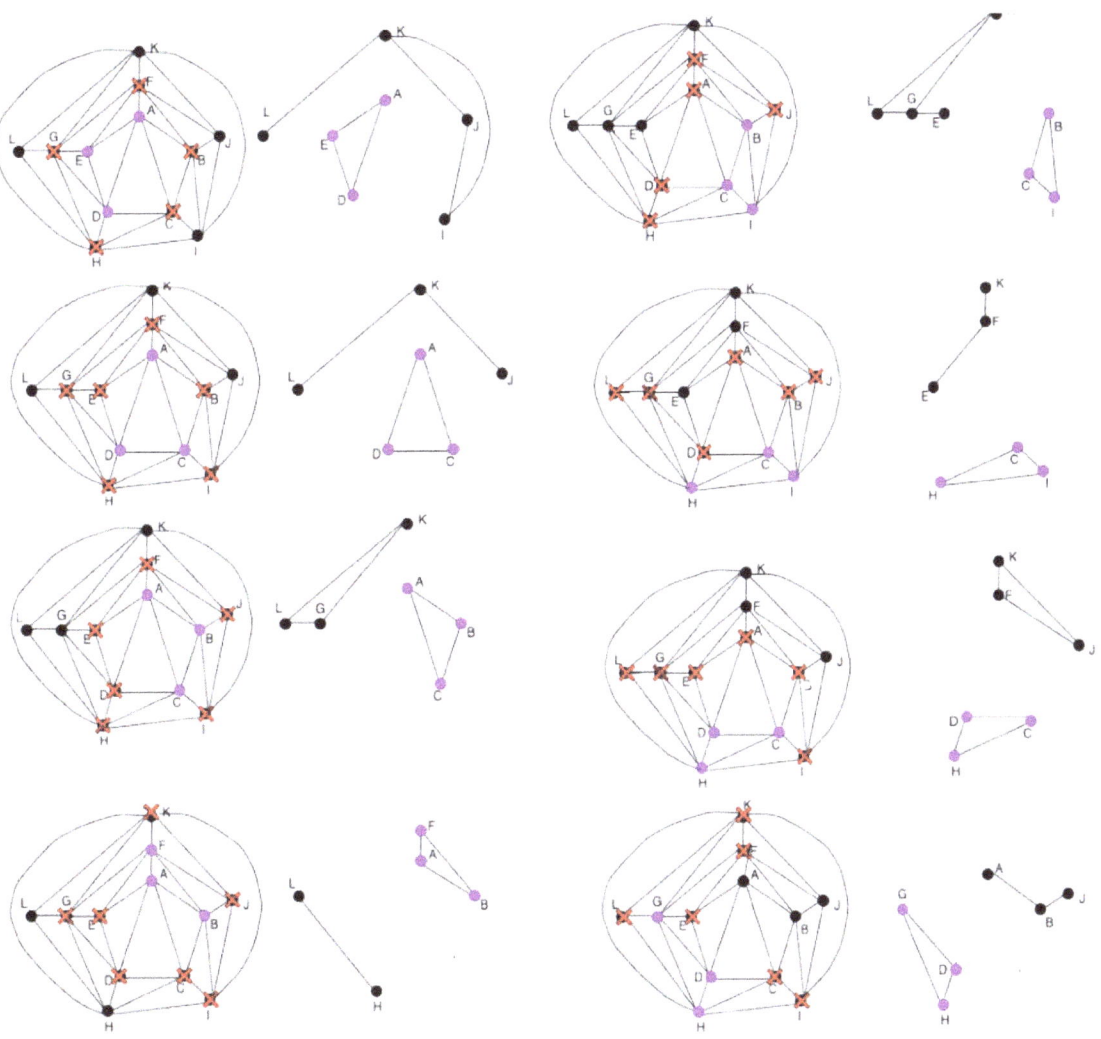

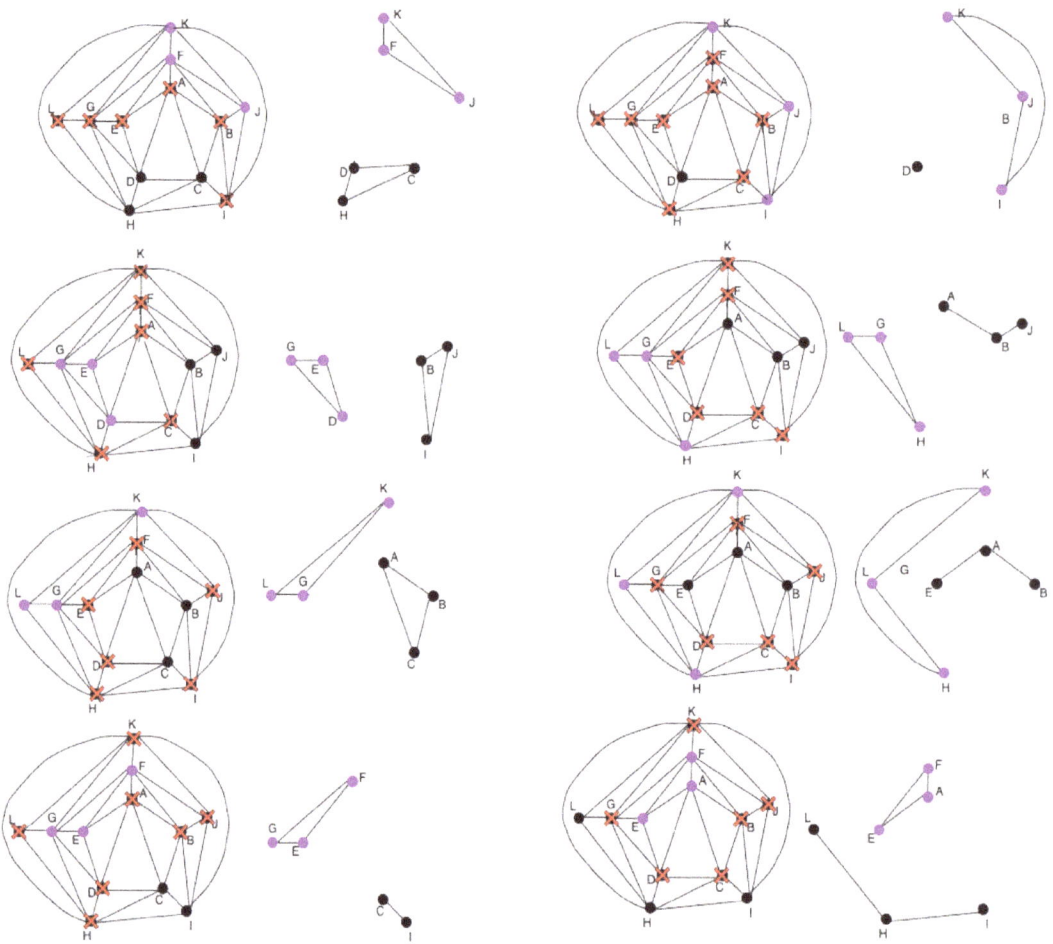

Figure 28: Isolating all triangles in the pentagonal structure

Again, as shown above, the pentagonal structure requires the removal of at least four vertices to isolate a triangle.

Diagonal Chromatic Number

In this paper, we will focus on the structures of plane triangulations of diameter three and one kind of coloring: diagonal coloring. Precisely speaking, we will show the following theorem:

Sierra's Theorem: *If G is a plane triangulation with diameter three, then $\chi_d(G) = 6$ when G is a specific graph with base shape is a triangle, square, or pentagon on twelve vertices.*

The triangle, square and pentagon structures had very distinct and similar outcomes. The illustrations below are the diagonal coloring of the triangular, square, and pentagonal figures.

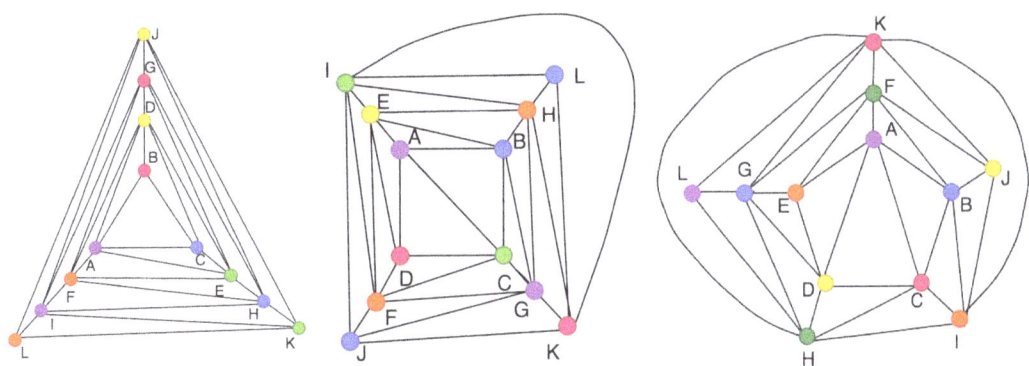

Figure 29: Diagonal Coloring of the triangle, square, and pentagon structures

The triangular structure had an interesting coloring pattern. By looking at the figure, we can see that two colors alternate alongside each side of the graph. For example, the vertices C, E, H, and K alternate between the colors green and blue, while vertices A, F, I and L alternate between the colors orange and purple. This pattern continues as long as the graph is expanded in the same form.

The pentagonal and square figures did not have any coloring pattern like the triangular structure. An interesting characteristic of the colored graphs was the distance between vertices with identical colors. The distance between the vertices with the same colors is at least 2. For example, the distance between the two purple vertices on the pentagonal graph is three and the distance between the two green vertices is two. On the square structure, the distance between the two orange vertices is two, while the distance between the two green vertices is three.

Although these structures were developed with different base shapes and expanded in a similar way, the diagonal chromatic number for all of the graphs described above is six.

Diagonal Coloring of The Triangle

We will now devote this section to explain the diagonal coloring of the triangle structure. Begin by coloring the inner triangle. We'll assign vertices A, B and C three unique colors because they are adjacent to each other.

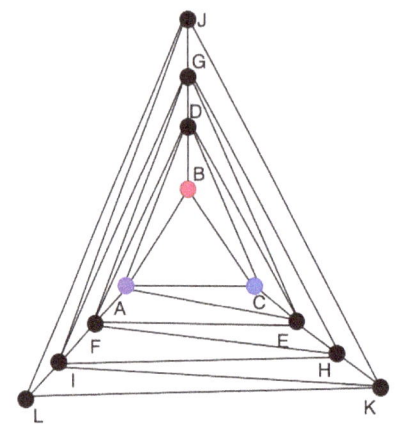

Figure 30: Diagonal coloring of vertices A, B and C

An additional color will need to be added in order to color vertex D, since vertex D is adjacent to vertices A, B, and C. As shown below vertex D will be of color yellow. Now we have four colors in our list of colors for this graph.

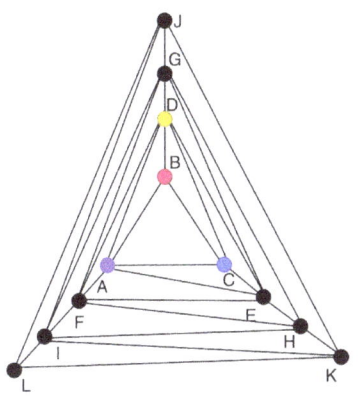

Figure 31: Diagonal coloring of vertex D

Vertex F cannot be of color purple, or yellow because it shares and edge with vertices A and D. In addition to purple and yellow, vertex F cannot be of color pink or blue because vertices B and C are diagonally adjacent to F. This means that vertex F will force a new color to our current list of colors. Vertex F will be colored orange.

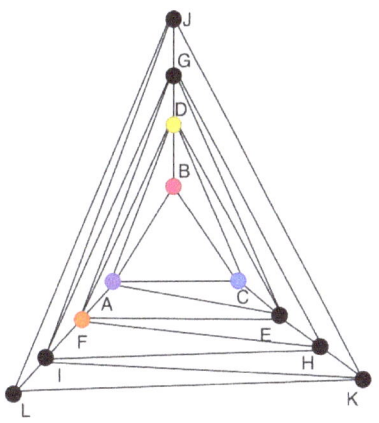

Figure 32: Diagonal coloring of vertex F

Similar to vertex F, vertex E cannot be of color orange, blue, purple, or yellow because it shares an edge with vertices A, C, D, and F. Vertex E cannot be of color pink or yellow because vertices B and D are diagonally adjacent to E. This implies an additional color must be added to our list of colors, which will be our sixth color. Vertex E will be of color green.

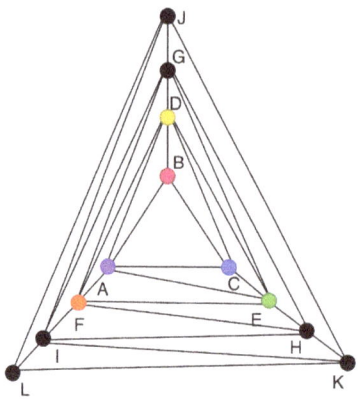

Figure 33: Diagonal coloring of vertex E

Next, we will color vertex G by first considering the colors that violate the definition of diagonal coloring. Vertex G cannot be yellow, orange or green because it is adjacent to vertices D, F, and E. However, since vertices A and B are neither diagonally adjacent or adjacent to vertex G, it can be either or those colors. Although we have two coloring options for vertex G, we need to strategically color the vertex such that no unnecessary color is added. Remember the goal is to use the minimum number of colors to color the graph. Thus, we will consider both colors.

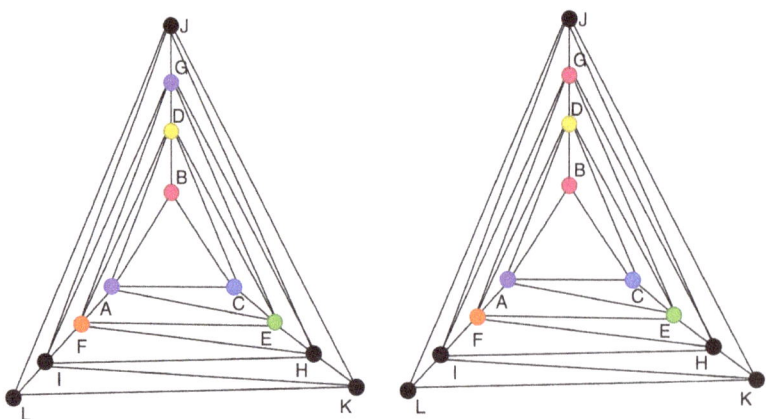

Figure 34: Diagonal coloring of vertex G. Triangular Structure A (left) and Triangular Structure B (right)

We will now begin to eliminate the colors that vertex I cannot be colored with. Since, vertex I is adjacent to vertices G, and F, vertex I in the triangular structure A cannot be of color purple or orange. Similarly, vertex I in the triangular structure B cannot be of color pink or orange. In both structures vertex I is diagonally adjacent to vertices D and E, which implies that it cannot be yellow or green. This means that in the triangular structure A, vertex I can be blue, or pink, while in the triangular structure B, vertex I can be blue or purple. Since we did not have to add a new color to our list, all of the graphs below are valid coloring for the triangular figure.

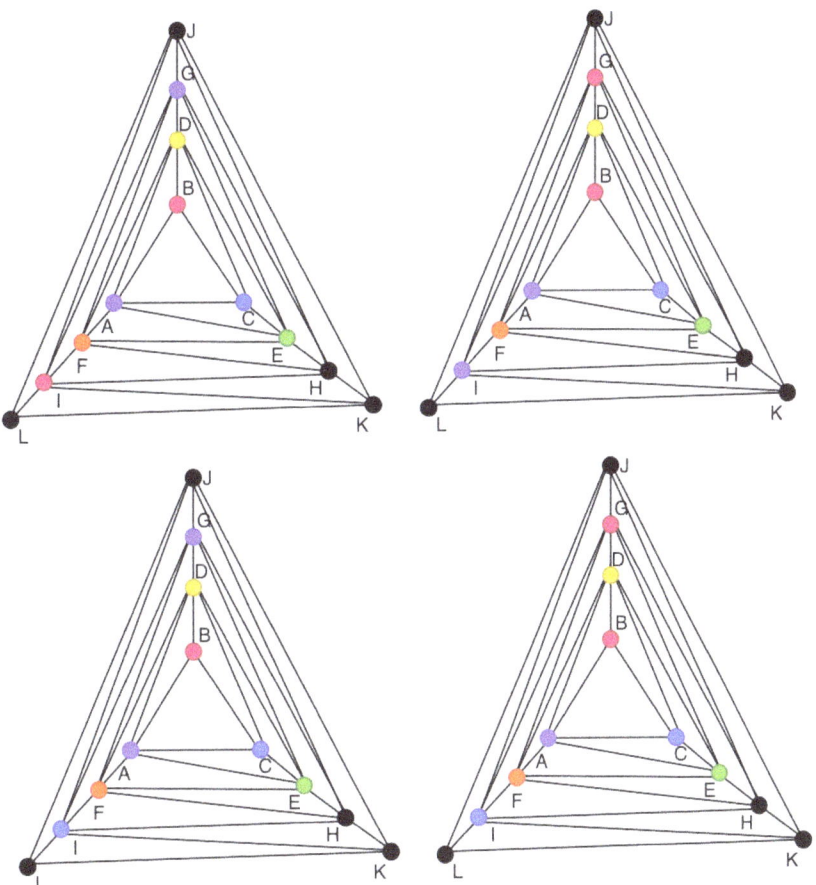

Figure 35: Diagonal coloring of vertex I. Triangular Structures A (top left), B (top right), C (bottom left), D (bottom right)

Now we must consider the possible colors for vertex H, by process of elimination. In triangular structure A and B, vertex H can only be blue. In the triangular structure C, vertex H can only be pink, and in structure D we would need to add an additional color to our list. In structure D, vertex H shares and edge with vertices I, E, F and G, which eliminate the possibility of it being colors blue, green, orange, and pink. In this structure vertex H is diagonally adjacent to vertices, D and A, which means that it cannot be yellow or purple. We can eliminate this graph because it is possible to color the graph with fewer colors like graph A, B and C.

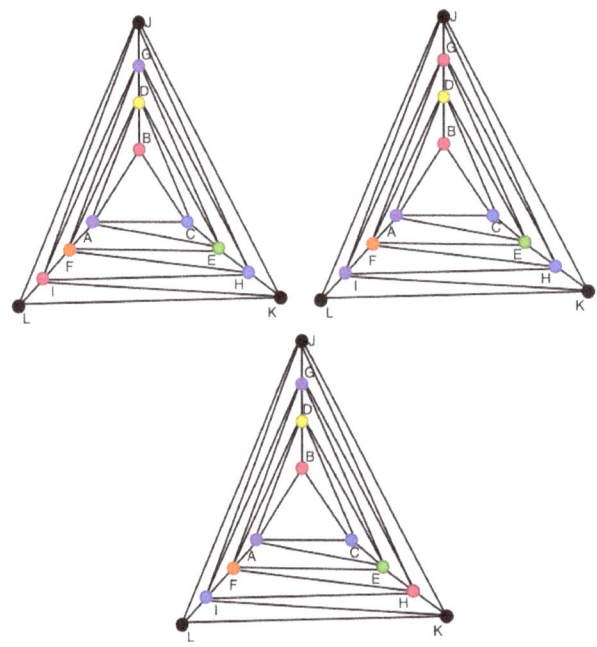

Figure 36: Diagonal coloring of vertex H. Triangular Structures A (top left), B(top right), C (bottom)

Vertex J is adjacent to vertices G, H and I. Vertex J is diagonally adjacent to vertices E and F. Surprisingly, the only color option for vertex J in triangular structure A, B and C is yellow.

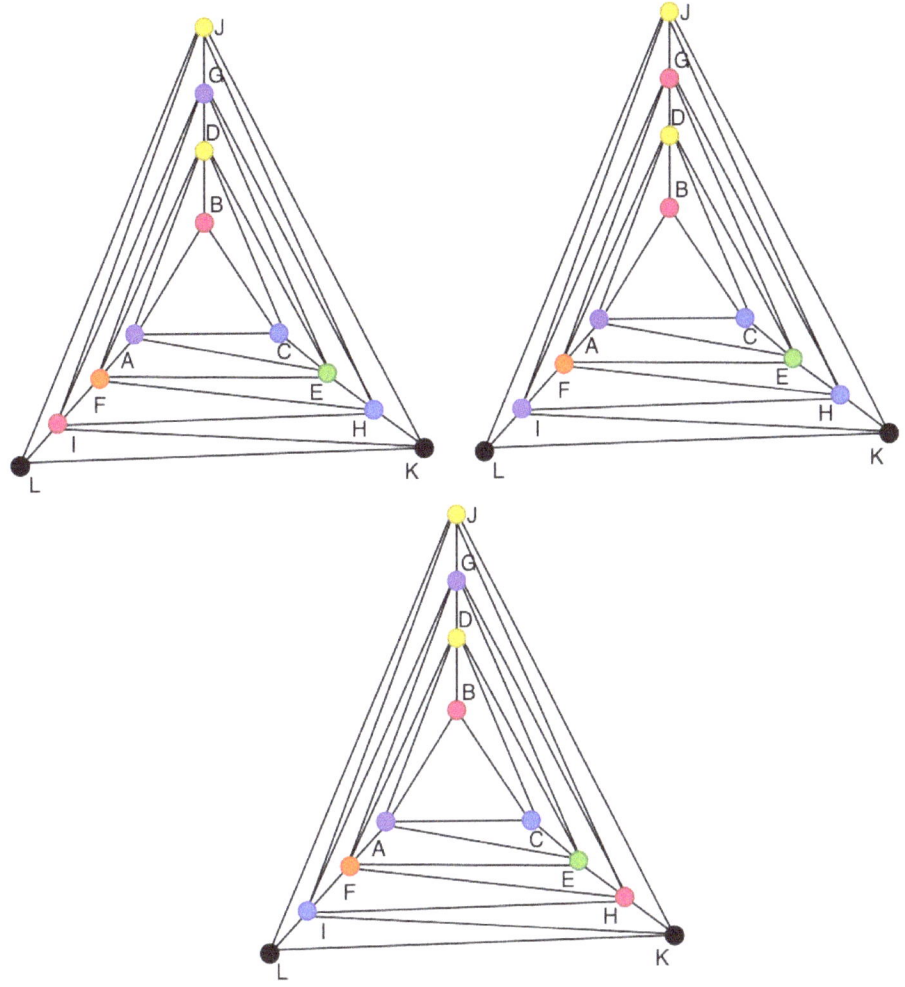

Figure 37: Diagonal coloring of vertex J. Triangular Structures A (top left), B(top right), C (bottom)

Vertex L shares and edge with colored vertices I, and J. Vertex L also is diagonally adjacent to

vertices H and G. In the triangular structure A, B and C, vertex L can be of color orange or

green.

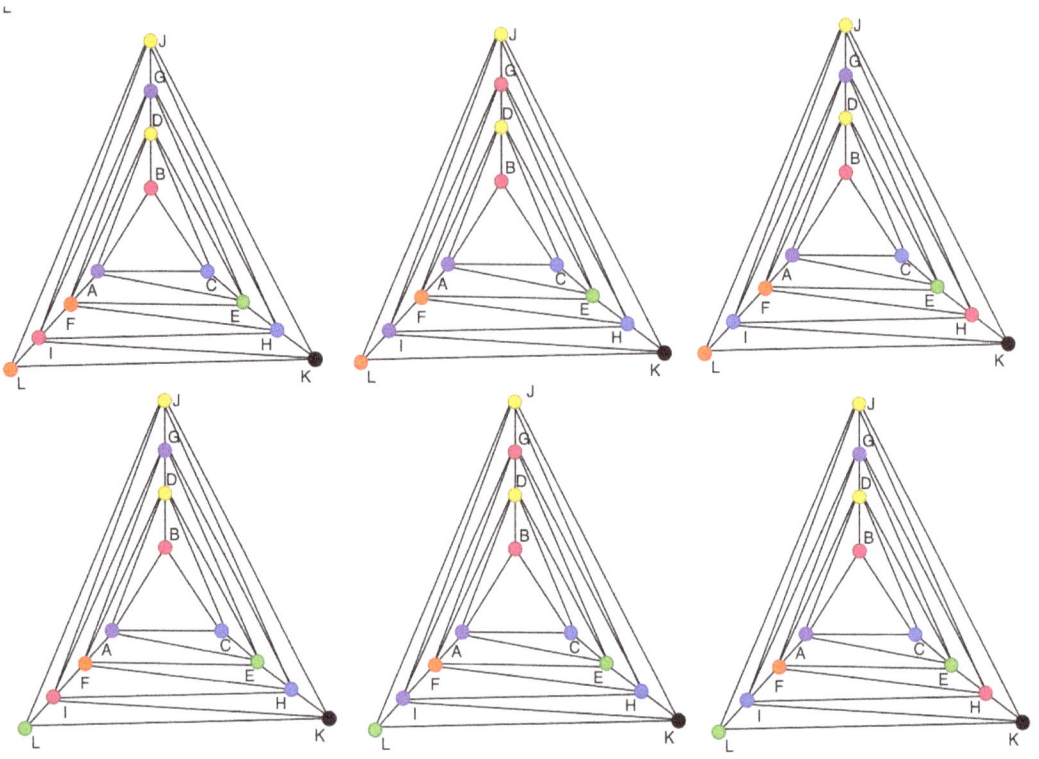

Figure 38: Diagonal coloring of vertex L. Triangular Structures A (top left), B (top middle), C (top right), D (bottom left), E (bottom middle), F (bottom right)

Lastly, we will determine the vertex color for vertex K. Vertex K is adjacent to vertices L, I, H, and J. It is also diagonally adjacent to vertices F and G. In triangular structures A, B and C, vertex K can only be of color green, because the vertices diagonally adjacent and adjacent to vertex K are of colors orange, pink, blue, yellow or purple. In order to color the triangular structures D, E and F an additional color must be added to our list of colors because the vertices diagonally adjacent and adjacent to vertex K are of colors green, pink, orange, blue, yellow, and purple. Since, we can color the graph with fewer colors we can eliminate graphs D, E and F. The proper coloring of the triangular structures is pictured below.

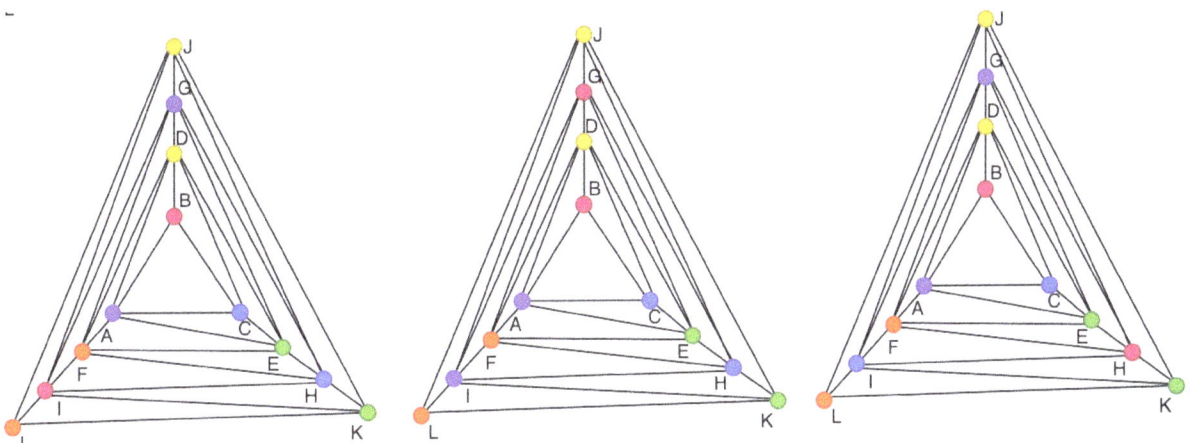

Figure 39: Valid diagonal coloring of the triangle structure

The diagonal chromatic number of the triangular structure is six. The list of colors are orange, blue, green, pink, purple, and yellow.

Diagonal Coloring of The Square

This section is devoted to explaining the diagonal coloring of the square structure.

First, we will begin by assigning vertex A and B different colors because these vertices are adjacent to each other.

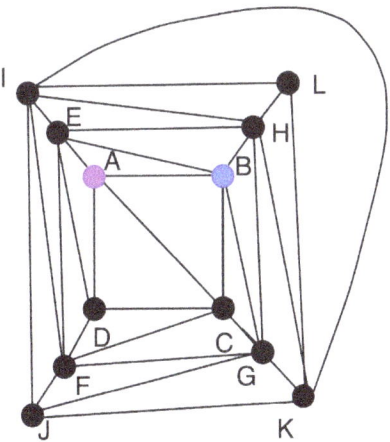

Figure 40: Diagonal coloring of vertices A and B

Vertices C and D will force two additional colors to be added to the current list of colors because these vertices are not only adjacent to each other, but also, they are either diagonally adjacent or adjacent to vertices A and B.

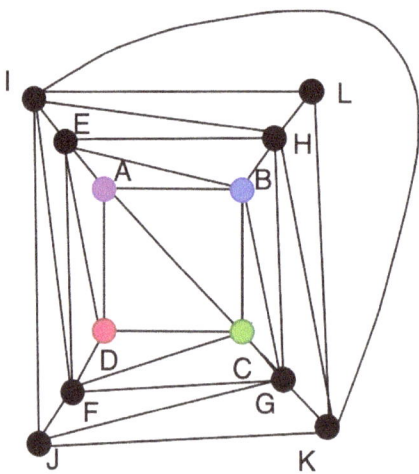

Figure 41: Diagonal coloring of vertices C and D

Vertex E will also force a new color in our list of colors due to the fact that it shares edges with vertices A, B, D and it is also diagonally adjacent to vertex C. Vertex E will be yellow.

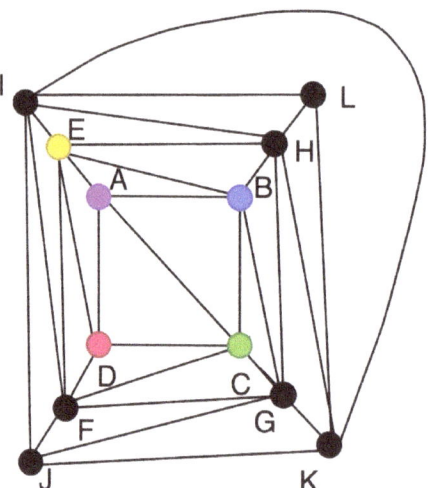

Figure 42: Diagonal coloring of vertex E

Similar to vertex E, Vertex F will also need a color that is not present on the graph because it is either adjacent or diagonally adjacent to vertices, A, B, C, D, and E.

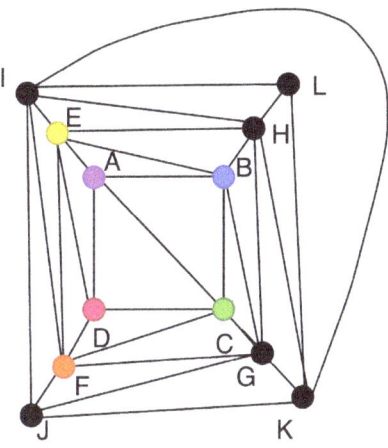

Figure 43: Diagonal coloring of vertex F

Vertex G is adjacent and diagonally adjacent to vertices F, C, D, B and E. The only possible

color vertex G can be is purple.

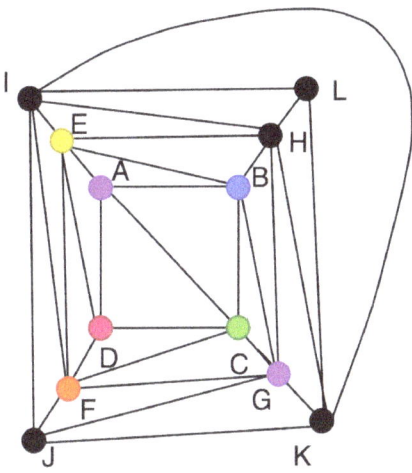

Figure 44: Diagonal coloring of vertex G

Vertex H is diagonally adjacent to vertices A and C. It is also adjacent to vertices E, B and G.

This implies that vertex H cannot be of color yellow, blue, purple, or green. The two possible

colors for vertex H is orange or pink.

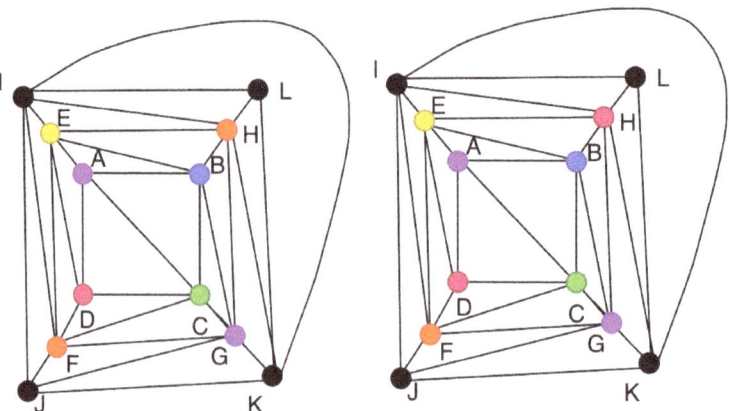

Figure 45: Diagonal coloring of vertex H. Square Structure A (left), B (right)

Vertex I is adjacent to vertices E, H, and F. It is also diagonally adjacent to vertices B and D. In the both square structure A and B, vertex I can only be of color green.

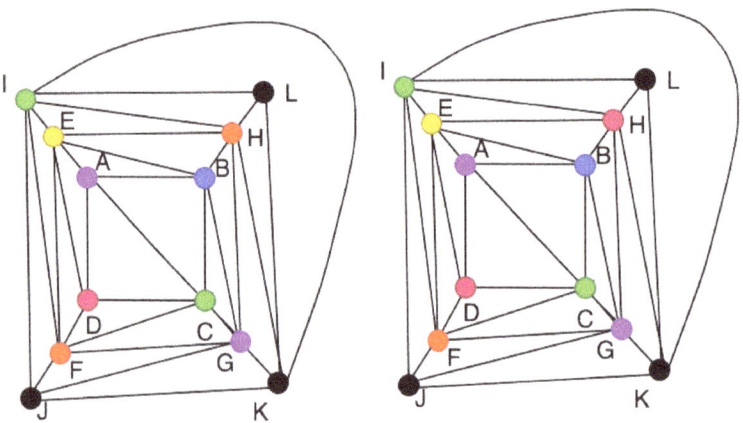

Figure 46: Diagonal coloring of vertex I. Square Structure A (left), B (right)

Vertex J shares an edge with vertices I, F, and G. It also is diagonally adjacent to vertices E, C and H. This means that for square structure A vertex J cannot be of color green, yellow, orange, or purple. The only possible colors that vertex J in square structure A can be either blue or pink. For square structure B, vertex J cannot be colored green, yellow, orange, purple, or pink, so the only color it can be is blue.

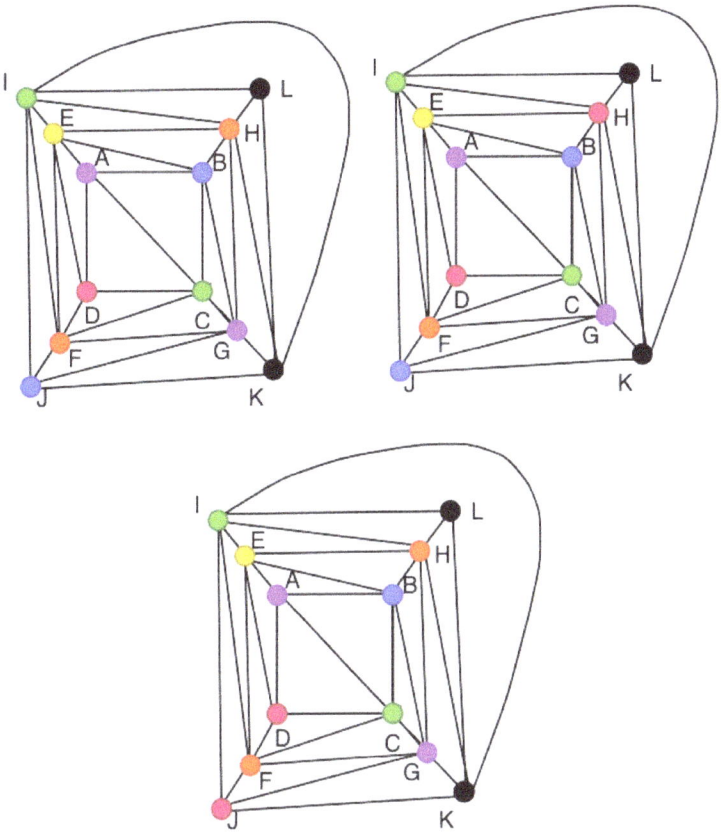

Figure 47: Diagonal coloring of vertex J. Square Structure A (top left), B (top right), C (bottom middle)

Vertex K is adjacent to vertices J, G, H and I. Vertex K is also diagonally adjacent to vertices E, F, B and I. Vertex K in the square structure A in figure 47. can only be of color pink. In order to color vertex K in both square structure B and C, we would need to add an additional color to our coloring lists. Thus, we can remove these graphs due to the fact that we can color the graph with less colors like the square structure A, which only requires six colors.

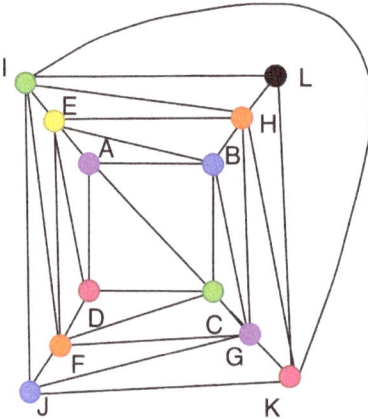

Figure 48: Diagonal coloring of vertex K

Lastly vertex L, cannot be of color yellow, green orange, pink or purple because the vertices that are diagonally adjacent or adjacent to L hold those colors. The only possible color it can be is blue.

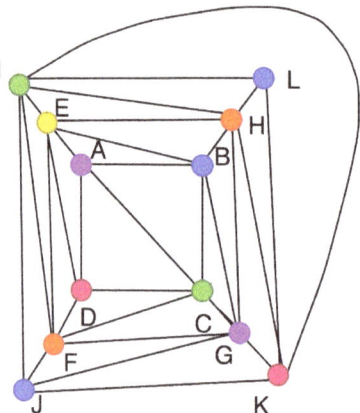

Figure 49: Diagonal coloring of vertex L

The diagonal chromatic number of the square structure is six. The colors are orange, blue, green, pink, purple, and yellow.

Diagonal Coloring of The Pentagon

This section is devoted to explaining the diagonal coloring of the pentagonal structure.

First, we will begin by assigning vertex A and B different colors because these vertices are adjacent to each other.

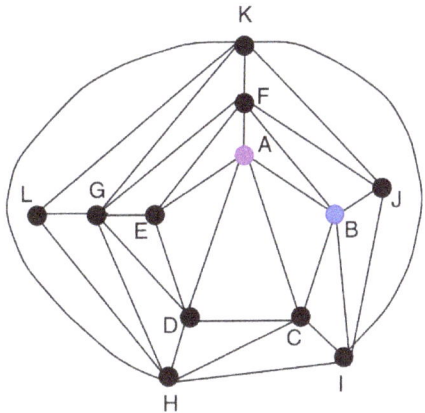

Figure 50: Diagonal coloring of vertices A and B

Vertex C will force a new color to be added to our list of colors because vertex C shares an edge

with vertices A and B.

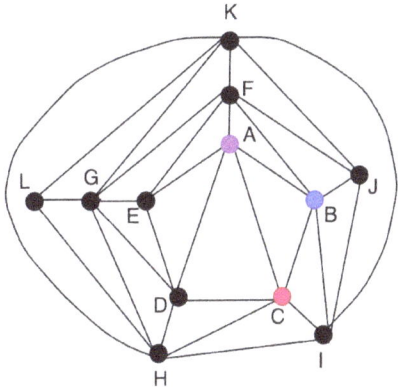

Figure 51: Diagonal coloring of vertex C

Vertex D will also force a new color to be added to our coloring list. It shares an edge with

vertices A and C, which implies that it cannot be of color purple or pink. Vertex D is also

diagonally adjacent to vertex B, which means that it cannot be of color blue. Vertex D will be of

color yellow.

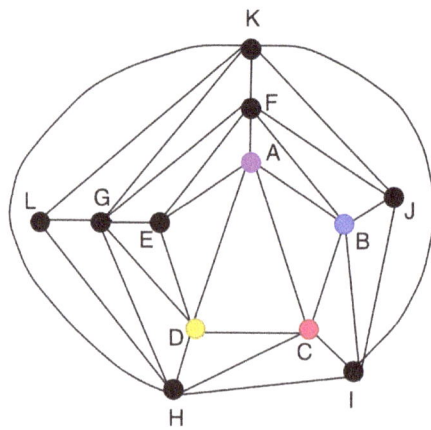

Figure 52: Diagonal coloring of vertex D

Vertex E is adjacent to vertices A and D, which means it cannot be of color purple or yellow. Vertices B and C are diagonally adjacent to vertex E, so it cannot be of color pink or blue. A new color must be added to our list of colors since the colors on the graph belong to vertices that are either adjacent or diagonally adjacent to vertex E.

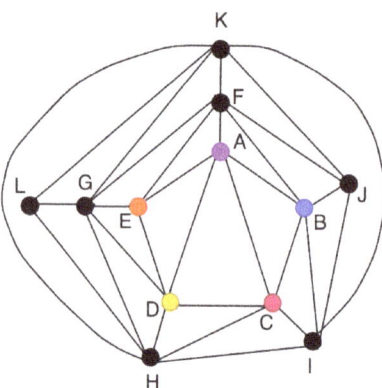

Figure 53: Diagonal coloring of vertex E

Vertex F shares an edge with vertices A, B and E. It is diagonally adjacent to vertex D, and C. Since, vertex F is connected to all of the vertices that are colored, we will need to add another color to our list of colors. Vertex F will be of color green. We now have the colors purple, blue, pink, yellow, orange, and green in our list of colors.

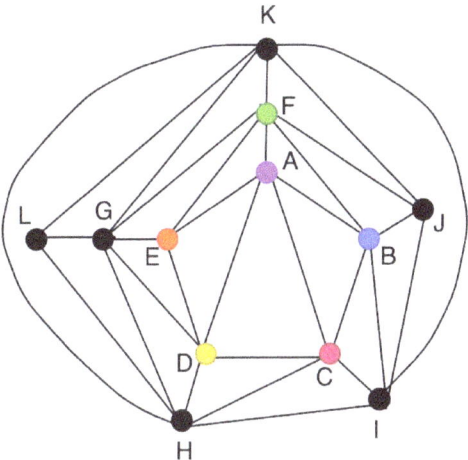

Figure 54: Diagonal coloring of vertex F

We must now consider the possible colors for vertex G, by utilizing process of elimination. Since vertex G is adjacent to vertices F, E and D, that implies it cannot be of color orange, green, or yellow. Vertices A, and C are diagonally adjacent to G, which also means that colors pink and purple are excluded from the possible colors for vertex G. The only possibly color that G can be is blue.

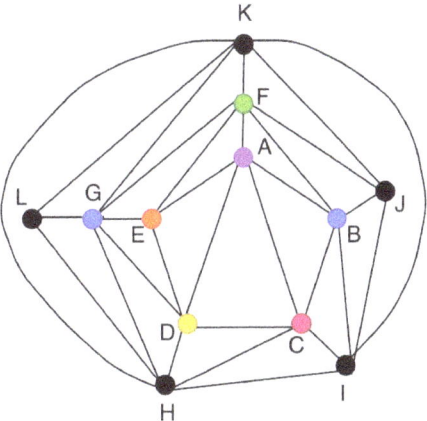

Figure 55: Diagonal coloring of vertex G

The vertices diagonally adjacent and adjacent to vertex H are A, B, C, D, E, and G. This implies that vertex H cannot be colored purple, blue, pink, yellow, or orange. The only possible color vertex H can have is green since vertex F is neither adjacent or diagonally adjacent to H.

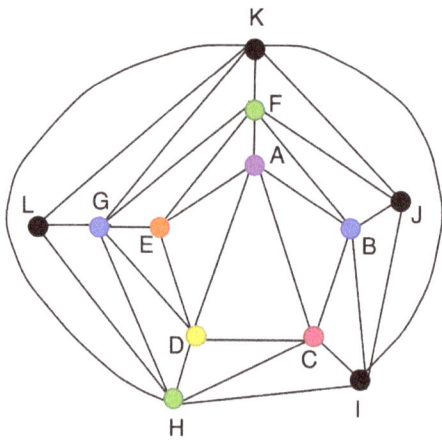

Figure 56: Diagonal coloring of vertex H

Vertex I shares an edge with vertices B, C, and H, which means that it cannot be of color green, pink or blue. The vertices diagonally adjacent to vertex I are A, D and F, which implies that it cannot be of colors purple, yellow or green. The only possible color vertex I can be is orange.

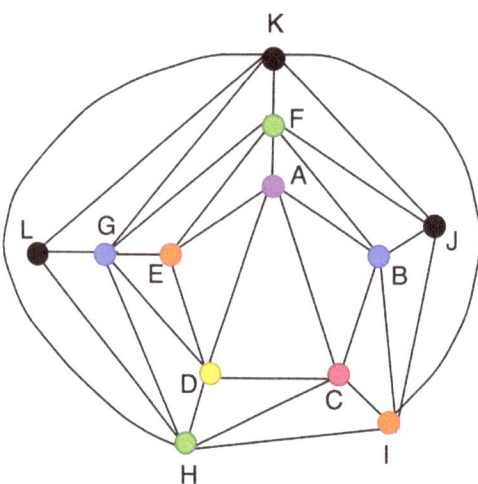

Figure 57: Diagonal coloring of vertex I

Vertex J is adjacent to vertices B, F and I, which means that J cannot be of colors green, blue or orange. Vertices C, A and G are diagonally adjacent to vertex J, so it cannot be of color pink, or purple. The only possible color that vertex J can be is yellow.

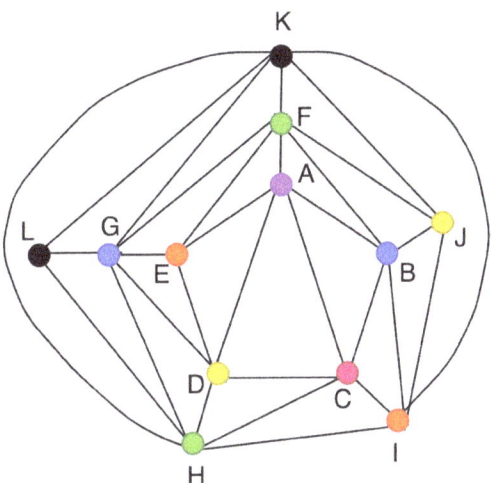

Figure 58: Diagonal coloring of vertex J

Vertex K is adjacent to vertices G, F, J, I and H. It is also diagonally adjacent to vertices G, B, E, and H. This means that vertex K cannot be of color blue, green, yellow, or orange. The two possible colors the K can hold are pink or purple. This is the first vertex in the graph to have two possible options for coloring.

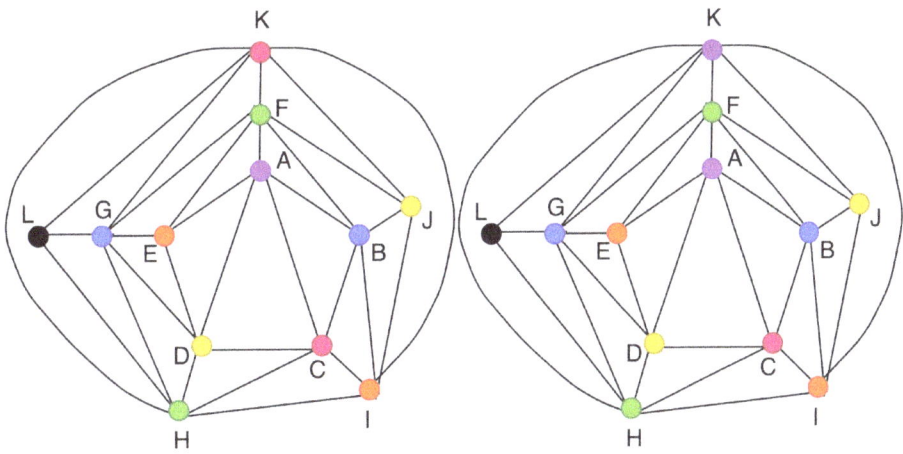

Figure 59: Diagonal coloring of vertex K. Pentagonal Structure A (left), B (right)

The last vertex that we will color is vertex L. Vertex L is adjacent to vertices G, H and K.

Vertices F, and D are diagonally adjacent to L. In the pentagonal structure A, vertex L cannot be of colors pink, green, blue, or yellow, so the only color selections for this vertex are purple or orange. In the pentagonal structure B, vertex L cannot be of colors purple, green, blue, or yellow, which means that the only possible colors this vertex can be is of color pink or orange. Below are the possible coloring for the pentagonal structure.

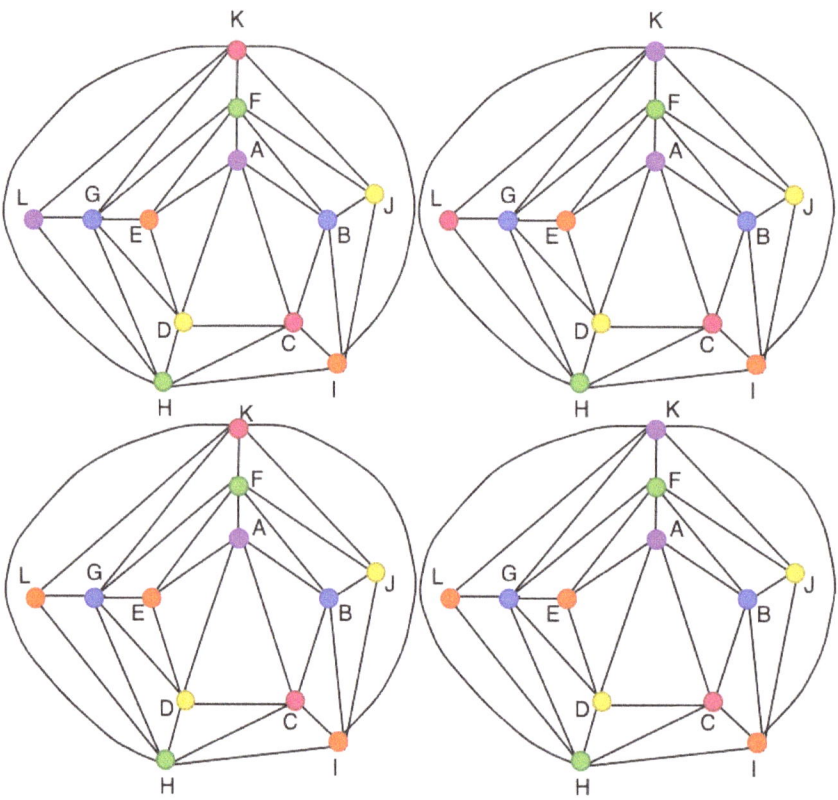

Figure 60: Diagonal coloring of the pentagonal structure

There are four possible ways to color the pentagonal structure such that the diagonal chromatic number of the graph remains six. Thus, we have proven Sierra's Theorem.

Conclusion

In this paper we discussed the diagonal chromatic number of maximal planar graphs of diameter less than four and a maximum of twelve vertices. The triangular, square, and pentagonal structure all had a diagonal chromatic number of six. Although the types of graphs that were discussed in this paper were specific in the way that the graphs were expanded, the number of vertices, and the diameter, this topic can be further explored by others. In continuation to this research, it would be interesting to further investigate all possible graphs of diameter of at least three because there could be a possible relationship between all possible graphs. Exploring graphs of various diameter could also serve as an extension to this research. Graph coloring will continue to be an asset to mathematics whether you're exploring the four-color theorem, diagonal coloring, or other forms of graph coloring.

Bibliography

Chartrand, Gary and Ping Zhang. *A First Course In Graph Theory*. Mineola, New York: Dover

      Publications, INC., 2012.

Huang, Danjun, et al. "List Coloring and." *Applied Mathematics and Computation* (2019).